A Straightforward Guide to

CYBER SECURITY FOR SMALL TO MEDIUM SIZE BUSINESS

A Straightforward Guide to

CYBER SECURITY
For Small to Medium Size Business

HOW TO ENSURE YOUR BUSINESS
IS PREPARED TO COMBAT A CYBER ATTACK

David Marsh

Straightforward Publishing
www.straightforwardco.co.uk

Straightforward Guides

© Straightforward Co Ltd 2022

All rights reserved. No part of this publication may be reproduced in a Retrieval system or transmitted by any means, electronic or mechanical, Photocopying or otherwise, without the prior permission of the copyright holder.

British Library Cataloguing in Publication Data.
A catalogue record is available for this book from the British library.

ISBN: 978-1-913776-99-2

Printed by 4edge www.4edge.co.uk
Cover design by BW-Studio Derby
Typeset in the UK by Frabjous Books

Whilst every effort has been taken to ensure that the information Contained within this book is accurate at the time of going to press, the authors and publishers recognise that the information can become out of date. The book is therefore sold on the condition that no responsibility for errors or omissions is assumed. The author and the publisher cannot be held liable for any information contained within.

Contents

Introduction   1

PART 1
AN OVERVIEW OF CYBER SECURITY

Chapter 1: Improving Cyber Security Generally – An Overview   9
Passive and Active security cyber security   9
What steps to take to improve your cybersecurity   10
One final word   16

PART 2
THE BASICS – BACKING UP DATA!

Chapter 2: The Importance of Backing up data   19
Definition of data backup   19
Identification of data that you will need to back up   20
Using the cloud to back up   21

Costs of using the cloud 22
Advantages of using the cloud 23
Disadvantages 24
Managed services 25
Main points from Chapter 2 28

PART 3
THE IMPACT OF MALICIOUS SOFTWARE

Chapter 3: Protecting Your Business from Malicious Software — 33

How to protect yourself against malicious code 33
What you need to know about antivirus software 36
Recovering if a victim of malicious code 38
Main points from chapter 3 40

PART 4
PROTECTING EMAILS

Chapter 4: Cyber Security and Emails — 45

Why can email attachments be dangerous? 45
Steps to be taken to protect self and others 46
Main points from chapter 4 49

Contents

PART 5
MANAGING PASSWORDS

Chapter 5: The Use of Effective Passwords to Protect Data 53

Why you need strong passwords 53
How to choose good passwords 55
   Length and complexity 56
   Do's and don'ts 57
How to protect your passwords 58
   Help your staff cope with password overload 58
Password managers 59
   Don't forget security basics 59
Main points from Chapter 5 61

PART 6
UTILISING FIREWALLS

Chapter 6: Using Firewalls to Protect Your System 65

Understanding Firewalls 65
What do firewalls do? 65
What type of firewall is best? 66
What configuration settings to apply 67
The most appropriate firewalls for business 68
Firewall terms to know 77
Main points from chapter 6 overleaf 82

PART 7
PHISHING AND OTHER ATTACKS

Chapter 7: Avoiding Phishing and Other Attacks — 85
What is a social engineering attack? — 85
What is a phishing attack? — 86
What is a vishing attack? — 86
What is a smishing attack? — 87
Common indicators of phishing attempts — 87
How do you avoid being a victim? — 89
What to do if you think you are a victim — 90
Main points from Chapter 7 — 92

PART 8
LOOKING AFTER MOBILES, TABLETS AND OTHER DEVICES

Chapter 8: Protecting Mobiles, Tablets and other Portable Devices — 97
Keep your Company Devices Safe from Threats — 97
Establish a Mobile Device Security Policy — 98
Establish a Bring Your Own Device Policy — 98
Keep Devices Updated with Current Software — 99
Backup Device Content on a Regular Basis — 100
Choose Passwords Carefully — 100
Mobile security software — 101

Contents

Mobile security software features	102
Main Points from Chapter 8	106
Conclusion and Summary of the book	109
Useful websites	111
Glossary of cyber security terms	113
Index	123

Introduction

The changing face of business

Technology is at the forefront of business innovation these days with a lot of interconnectivity across devices and cloud services. Cyber security used to be as simple as ensuring your IT network had up-to-date virus protection, but, as we shall see, it's far more complicated in the current climate.

As older readers might remember, before computers came on the scene, businesses stored all of their data in physical files stored in filing cabinets. The only way they could be accessed was by physical theft, either by outsiders or, in some cases, insiders.

As the use of information technology has evolved, and offices and organisations became paperless, so has it become easier for outsiders to access data. This doesn't require a physical break-in. It requires sophisticated use of information technology to 'Hack' systems and either steal data or plant a virus to hold companies to ransom, as is increasingly becoming the case. The only real way to combat this, for businesses of all sizes, is by employing ever more sophisticated Cyber-

Security policy and practice across business to prevent unauthorized access to systems.

A business might start in the most humble settings, as do many start-ups. However, in many cases the use of computers will grow as a business grows and the amount of data that is recorded also grows. It is very important to be aware from the start of the need to implement a sophisticated cyber security strategy to protect data from being stolen

Defining cyber security
To put it simply, cybersecurity is the art of protecting computer networks, devices, and data from unauthorized access or criminal use and the practice of ensuring confidentiality, integrity, and availability of information.

Ransomware gangs have increasingly focused on high-profile targets like large corporations and government institutions in the past year, according to Europol's *Internet Organised Crime Threat Assessment (IOCTA) 2021*. However, small to medium size businesses are also very much affected. The report, which offers insights into current cybercrime trends in Europe, revealed that ransomware actors have taken advantage of widespread home-working to launch more sophisticated and targeted attacks.

The report also highlighted the growing use of multi-

Introduction

layered methods to extort service providers, financial institutions and businesses, such as DDOS attacks (Denial of Service Attack). A DDOS Attack is a malicious attempt to disrupt normal traffic of a targeted server, service or network by overwhelming the target or its surrounding infrastructure with a flood of Internet traffic.

Additionally, they observed that cyber-criminals have increasingly recognized the potential to attack a large number of organizations via supply chain attacks, often targeting the 'weakest link.'

Research has also shown that more than eight in ten mid-sized businesses experienced fraud in 2021, up from 60 per cent in the previous year. Cyber-attacks were one of the most prevalent types of fraud. Much of this fraud has been enabled by more and more people working from home. Cyber security wasn't able to keep up with the sudden transition in working practices.

Other notable trends in the past year included fraudsters continuing to exploit the COVID-19 crisis and increasingly target online shopping to scam victims. There has also been an evolution in mobile malware, with cyber-criminals trying to find ways to circumvent additional security measures such as two-factor authentication, according to the report.

This is further evidence of how much of a threat ransom attacks pose to businesses, including those that go beyond ransomware. It's no coincidence that the number of DDOS attacks has quadrupled in the last year. Using rapid-fire attacks, averaging just six minutes, cyber-criminals demonstrate their capabilities to businesses before sending an extortion demand, threatening much larger attacks if payments aren't made.

Hackers are carrying out ransom attacks because they are one of the fastest ways to big profits, and their tactics go beyond just using malware. Businesses need to have proper cyber-resiliency strategies in place so that no matter what sort of ransom attack comes their way, the impact is minimized and operations can continue.

In summary

It seems now that everything we do in a business and personal sense, relies on computers and the internet, communication (e.g., email, smartphones, tablets), entertainment (e.g., interactive video games, social media, apps), transportation (e.g., navigation systems), shopping (e.g., online shopping, credit cards), medicine (e.g., medical equipment, medical records), and the list goes on. Stop to think, how much of your daily life, both in a business setting and also personally, relies on technology? How much of your personal information is stored either on your own computer, smartphone, tablet or on someone else's system? It's frightening to think.

Introduction

However, one thing is absolutely certain, and it is what this brief book is about, is that protection is the key. Protection of your business data and protection of your personal data (the two are intertwined).

What are the risks of having poor cybersecurity?
As has been pointed out, there are many risks associated with poor cyber security, some much more serious than others. Among these dangers, which we discuss and elaborate on throughout the book, are **malware** erasing your entire system (a nightmare), an attacker breaking into your system and altering files, an attacker using your computer to attack others, or an attacker stealing your credit card information and making unauthorized purchases.

There is no guarantee that even with the best precautions some of these things won't happen to you, but there are steps you can take to minimize the chances. These steps will be discussed as we go through the book, chapter by chapter.

Part 1

OVERVIEW OF CYBER SECURITY

Chapter 1

Improving Cyber Security Generally: An Overview

In this chapter, we look generally at ways to improve cyber security in your business. In subsequent chapters we will elaborate on each area.

Passive and Active cyber security
Passive security is what most businesses have in place already i.e. all the critical applications like Antivirus protection, network and endpoint security and all other defensive equipment and software which aims to continually protect an environment from outside threats. The question is: these components are essential, but how do you know if they are working?

This is where Active cyber security comes in. It's the offensive approach, purposefully identifying exploitable weaknesses. For example, penetration testing is designed to simulate a hacker who is attempting to bypass your defences and phishing simulations are designed to test and harden your weakest defences (human beings) by sending very convincing emails that appear real but include malicious links.

Both penetration testing and phishing simulations are highly effective, and don't impact business operations during tests i.e. no damage is done. These are just two basic examples of active engagement. There are many more security services like this which aim to thoroughly test your defences.

The goal is to identify every possible weak point; the longer term the campaign, the more effective it is. By applying both passive and active measures, a business can improve the reliability of both passive security solutions and the vigilance of team members all in one go.

What steps can you take to improve your cybersecurity?
Taking into account the above approaches, the very first step in protecting yourself and your business is to recognize the risks.

It helps to be familiar with the following terms to better understand the risks, as outlined below.

(There is a full glossary of terms at the end of the book).

Improving Cyber Security Generally: An Overview

- **Hacker, attacker, or intruder** – These terms are applied to the people who seek to exploit weaknesses in software and computer systems for their own gain. Although their intentions are sometimes benign and motivated by curiosity, their actions are typically in violation of the intended use of the systems they are exploiting. The results can range from mere mischief (creating a virus with no intentionally negative impact) to malicious activity (stealing or altering information).

- **Malicious code (see chapter 2)** – Malicious code (also called malware) is unwanted files or programs that can cause harm to a computer or compromise data stored on a computer. Various classifications of malicious code include viruses, worms, and Trojan horses.

Malicious code may have the following characteristics:

- It might require you to actually do something before it infects your computer. This action could be opening an email attachment or going to a particular webpage.
- Some forms of malware propagate without user intervention and typically start by

exploiting a software vulnerability. Once the victim computer has been infected, the malware will attempt to find and infect other computers. This malware can also propagate via email, websites, or network-based software.

- Some malware claims to be one thing, while in fact doing something different behind the scenes. For example, a program that claims it will speed up your computer may actually be sending confidential information to a remote intruder.

Vulnerabilities – Vulnerabilities are flaws in software, firmware, or hardware that can be exploited by an attacker to perform unauthorized actions in a system. They can be caused by software programming errors. Attackers take advantage of these errors to infect computers with malware or perform other malicious activity. To minimize the risks of cyberattacks, follow basic cybersecurity best practices:

- **Keep software up to date.** Install software patches so that attackers cannot take advantage of known problems or vulnerabilities. A software patch or fix is a **quick-repair job** for programming designed to resolve functionality issues, improve

Improving Cyber Security Generally: An Overview

security and add new features. Throughout its lifetime, software will run into problems called bugs. A patch is the immediate fix to those problems. Most software programs may have several patches after their initial release and usually update the version of the program when successfully installed. Software patches, when available, can generally be found at the software developer's website. Many operating systems offer automatic updates. If this option is available, you should enable it.

- **Run up-to-date antivirus software.** A reputable antivirus software application is an important protective measure against known malicious threats. It can automatically detect, quarantine, and remove various types of malware. Be sure to enable automatic virus definition updates to ensure maximum protection against the latest threats. Note: Because detection relies on signatures – known patterns that can identify code as malware – even the best antivirus will not provide adequate protections against new and advanced threats, such as zero-day exploits and polymorphic viruses.

- **Use strong passwords (see chapter 5).** Select passwords that will be difficult for attackers to guess, and use different passwords for different programs and devices. It is best to use long, strong passphrases or passwords that consist of at least 16 characters.

- **Change default usernames and passwords.** Default usernames and passwords are readily available to malicious actors. Change default passwords, as soon as possible, to a sufficiently strong and unique password.

- **Implement multi-factor authentication (MFA).** Authentication is a process used to validate a user's identity. Attackers commonly exploit weak authentication processes. MFA uses at least two identity components to authenticate a user's identity, minimizing the risk of a cyber-attacker gaining access to an account if they know the username and password.

- **Install a firewall (see chapter 6).** Firewalls may be able to prevent some types of attack vectors by blocking malicious traffic before it can enter a computer system, and by restricting unnecessary outbound communications. Some device operating systems include a firewall. Enable and properly configure the firewall as specified in the device or system owner's manual.

Improving Cyber Security Generally: An Overview

- **Be suspicious of unexpected emails (see chapter 4).** Phishing emails are currently one of the most prevalent risks to the average user. The goal of a phishing email is to gain information about you, steal money from you, or install malware on your device. Be suspicious of all unexpected emails.

Most importantly, and this is discussed in the next chapter, BACK UP YOUR DATA. Easy to forget to do but potentially lethal if you are a victim of a cyber attack.

One final word

This book is very much about DIY cyber security. It is not overly technical so as to baffle those without a high level of technical knowledge but is very much pitched at a middle level, presupposing a reasonable understanding of computerised systems.

It is, above all, about the steps that you can take to prevent unauthorized access to your systems, in whatever form. However, in case you think you would benefit from professional help, I have included websites that list the most recommended cyber security firms and also government organisations at the end of this book.

Part 2

THE BASICS
BACKING UP DATA!

Chapter 2

The Importance of Backing up Data

Definition of data backup

Data backup is the process of protecting data in case of a disaster, accident, or malicious action, by copying it from one location to another. Data is the lifeblood of any organisation, losing data can lead to serious damage and interrupt business operations.

If you give thought to how much you actually rely on data that is critical to your business, you will then realise that the last thing that you need is for you to lose that data. This could ruin you. On your computers, you will have customer details, quotes, orders and details of payments made and payments due.

It goes without saying that all businesses, large or small, need to carry out data backups regularly. These backups should be recent and can be quickly restored ensuring that your business can still function following the impact of malicious attacks, flood, fire, physical damage or theft.

Identification of data that you will need to back up
Any data that you need access to in order to maintain 'business as usual' needs backing up. Also, any personal data you hold, such as customer profiles, transactional details – or any legal or financial information – needs a secure storage and access solution. As for when (i.e. how often), that really depends on the kind of business you operate and the data it creates, stores and depends on.

Websites or web-connected internal systems that create, use and depend on continuously updated and/or personal and private data should back up almost continuously – in some cases every few seconds – to avoid potentially serious problems if IT fails. Purely information-sharing or entertainment websites that don't harvest personal or transactional details, or depend on up-to-the-second data, can comfortably backup once or twice a day. If in doubt, it's probably best to ask an IT expert for help choosing the backup strategy that's best for you.

Some of these solutions are discussed below.

In both scenarios, an incremental backup, whereby only new or altered data is stored (rather than all of it) can be a more efficient approach.

It is very important that you keep your backup separate from your computer. Ransomware (and other malware) can often move to attached storage automatically, which means that your backup will be compromised, leaving you with no backup at all. You should consider moving your backups to a different location, so fire or theft won't result in you losing everything. It is also important to ensure that access to data backups is restricted so that they are not accessible to staff (except for those who should be given access).

Cloud storage solutions are a cost effective and efficient way of achieving this.

Using the cloud to back up
A remote, online, or managed backup service, sometimes marketed as **cloud backup** or **backup-as-a-service**, is a service that provides users with a system for the backup, storage, and recovery of computer files. Online backup providers are companies that provide this type of service to end-users (or clients). Such backup services are considered a form of cloud computing.

Online backup systems are typically built for a client software program that runs on a given schedule. Some systems run once a day, usually at night while computers aren't in use.

Other newer cloud backup services run continuously

to capture changes to user systems nearly in real-time. The online backup system typically collects, compresses, encrypts, and transfers the data to the remote backup service provider's servers or off-site hardware.

There are many products on the market – all offering different feature sets, service levels, and types of encryption. Providers of this type of service frequently target specific market segments.

High-end Local Area Network (LAN) based backup systems may offer services such as Active Directory, Client Remote Control, or Open File Backups. Consumer online backup companies frequently have beta software offerings and/or free-trial backup services with fewer live support options.

*

Costs of using the cloud
Online backup services are usually priced as a function of the following things:

1. The total amount of data being backed up.
2. The total amount of data being restored.
3. The number of machines covered by the backup service.
4. The maximum number of versions of each file that are kept.

5. Data retention and archiving period options
6. Managed backups vs. Unmanaged backups
7. The level of service and features available

Some vendors limit the number of versions of a file that can be kept in the system. Some services omit this restriction and provide an unlimited number of versions. Add-on features (plug-ins), like the ability to back up currently open or locked files, are usually charged as an extra, but some services provide this built-in.

Most remote backup services reduce the amount of data to be sent over the wire by only backing up changed files. This approach to backing up means that the customers total stored data is reduced. Reducing the amount of data sent and also stored can be further drastically reduced by only transmitting the changed data bits by binary or block level incremental backups. Solutions that transmit only these changed binary data bits do not waste bandwidth by transmitting the same file data over and over again if only small amounts change.

Advantages of using the cloud
Remote backup has advantages over traditional backup methods:

- Remote backup does not require user intervention. The user does not have to change tapes, label CDs or perform other manual steps.

- Unlimited data retention (presuming the backup provider stays in business).

- Some remote backup services will work continuously, backing up files as they are changed.

- Most remote backup services will maintain a list of versions of your files.

- Most remote backup services will use a 128 - 2048 bit encryption to send data over unsecured links (e.g. internet).

- A few remote backup services can reduce backup by only transmitting changed data.

- Manage and secure digital data information.

Disadvantages
Remote backup has some disadvantages:

- Depending on the available network bandwidth, the restoration of data can be slow. Because data is stored offsite, the data must be recovered either via the Internet or via a disk shipped from the online backup service provider.

The Importance of Backing up Data

- Some backup service providers have no guarantee that stored data will be kept private.

- It is possible that a remote backup service provider could go out of business or be purchased, which may affect the accessibility of one's data or the cost to continue using the service.

- If the encryption password is lost, data recovery will be impossible. However, with managed services this should not be a problem.

- Residential broadband services often have monthly limits that preclude large backups. They are also usually asymmetric; the user-to-network link regularly used to store backups is much slower than the network-to-user link used only when data is restored.

- In terms of price, when looking at the raw cost of hard disks, remote backups cost about 1-20 times per GB what a local backup would.

Managed services
Some services provide expert backup management services as part of the overall offering.

These services typically include:

- Assistance configuring the initial backup
- Continuous monitoring of the backup processes on the client machines to ensure that backups actually happen
- Proactive alerting in the event that any backups fail
- Assistance in restoring and recovering data

Scheduled vs manual vs Event based backup

There are three distinct types of backup modes: Scheduled, Manual and Event-based.

- Scheduled Backup – data is backed up according to a fixed schedule.
- Manual Backup – backup of data is triggered by user input.
- Event-based Backup – backup of data is triggered by some computer events, e.g. database or application stoppage (cold backup).

Finally, you should ensure that your company policy clearly stipulates that a backup is carried out, in whichever form, regularly.

It is easy to forget to backup but it is crucial that you don't neglect it! This applies to all types of business.

The Importance of Backing up Data

For detailed guidance on cloud security, go to the National Cyber Security Centre:

https://www.ncsc.gov.uk/collection/cloud-security/implementing-the-cloud-security-principles

Now read the main points from Chapter 2 overleaf.

Main points from Chapter 2

BACKING UP DATA

- On your computers, if you run a business, you will have customer details, quotes, orders and details of payments made and payments due. It goes without saying that all businesses, large or small, need to carry out data backups regularly.

- These backups should be recent and can be quickly restored ensuring that your business can still function following the impact of flood, fire, physical damage or theft.

- It is very important that you keep your backup separate from your computer. Ransomware (and other malware) can often move to attached storage automatically, which means that your backup will be compromised, leaving you with no backup at all

- A remote, online, or managed backup service, sometimes marketed as cloud backup or backup-as-a-service, is a service that provides users with a system for the backup, storage, and recovery of computer files. Online backup providers are companies that provide this type of service to end users (or clients). Such backup services are considered a form of cloud computing.

- There are both advantages and disadvantages of using the cloud as backup. One advantage is that it takes the work out of your hands to a certain extent. However, there is a cost attached to using the cloud.

Part 3

THE IMPACT OF MALICIOUS SOFTWARE

Chapter 3

Protecting Your Business from Malicious Software

Malicious software (Malware) is one of the biggest cyber threats facing small to medium size business (and also individuals)

How can you protect yourself against malicious code? Following the below security practices can help you reduce the risks to your business associated with malicious code:

- **Install and maintain antivirus software.** Antivirus software recognizes malware and protects your computer against it. Installing antivirus software from a reputable vendor is an important step in preventing and detecting infections. Always visit vendor sites directly rather than clicking on advertisements or email links. Because attackers are continually creating new viruses and other forms of malicious code, it is important to keep your antivirus software up-to-date.

- **Use caution with links and attachments.** Take appropriate precautions when using email and web browsers to reduce the risk of an infection. Be wary of unsolicited email attachments and use caution when clicking on email links, even if they seem to come from people you know. (See chapter 4 for more about emails).

- **Block pop-up advertisements.** Pop-up blockers disable windows that could potentially contain malicious code. Most browsers have a free feature that can be enabled to block pop-up advertisements.

- **Use an account with limited permissions.** When navigating the web, it's a good security practice to use an account with limited permissions. If you do become infected, restricted permissions keep the malicious code from spreading and escalating to an administrative account.

- **Disable external media AutoRun and AutoPlay features.** Disabling AutoRun and AutoPlay features prevents external media infected with malicious code from automatically running on your computer.

- **Change your passwords (see chapter 5 for more about passwords).** If you believe your computer is infected, change your passwords. This includes any passwords for websites that may have been cached in your web browser. Create and use strong passwords, making them difficult for attackers to guess.

- **Keep software updated.** Install software patches on your computer so attackers do not take advantage of known vulnerabilities. *Consider enabling automatic updates, when available.*

- **Back up data.** As we have discussed in chapter 2, regularly back up your documents, photos, and important email messages to the cloud or to an external hard drive. In the event of an infection, your information will not be lost.

- **Install or enable a firewall (see chapter 6).** Firewalls can prevent some types of infection by blocking malicious traffic before it enters your computer. Some operating systems include a firewall; if the operating system you are using includes one, enable it.

- **Use anti-spyware tools.** Spyware is a common virus source, but you can minimize infections by using a program that identifies and removes spyware. Most antivirus software includes an anti-spyware option; ensure you enable it.

- **Monitor accounts.** Look for any unauthorized use of, or unusual activity on, your accounts – especially banking accounts. If you identify unauthorized or unusual activity, contact your account provider immediately.

- **Avoid using public Wi-Fi.** Unsecured public Wi-Fi may allow an attacker to intercept your device's network traffic and gain access to your personal information.

What do you need to know about antivirus software?
Antivirus software scans computer files and memory for patterns that indicate the possible presence of malicious code. You can perform antivirus scans automatically or manually.

- **Automatic scans** – Most antivirus software can scan specific files or directories automatically. New virus information is added frequently, so it is a good idea to take advantage of this option.

- **Manual scans** – If your antivirus software does not automatically scan new files, you should manually scan files and media you receive from an outside source before opening them, including email attachments, web downloads, CDs, DVDs, and USBs.

Although anti-virus software can be a powerful tool in helping protect your computer, it can sometimes induce problems by interfering with the performance of your computer. Too much antivirus software can affect your computer's performance and the software's effectiveness.

- **Investigate your options in advance.** Research available antivirus and anti-spyware software to determine the best choice for you. Consider the amount of malicious code the software recognizes and how frequently the virus definitions are updated. Also, check for known compatibility issues with other software you may be running on your computer.

- **Limit the number of programs you install.** Packages that incorporate both antivirus and anti-spyware capabilities together are now available. If you decide to choose separate programs, you only need one antivirus program and one anti-spyware program. Installing more programs increases your risk of problems.

There are many antivirus software program vendors, and deciding which one to choose can be confusing. Antivirus software programs all typically perform the same type of functions, so your decision may be based on recommendations, features, availability, or price. Regardless of which package you choose, installing any antivirus software will increase your level of protection.

Recovering if you become a victim of malicious code Ideally, your antivirus program will identify any malicious code on your computer and quarantine them so they no longer affect your system. You should also consider these additional steps:

- **Minimise the damage.** If you have access to an information technology (IT) department or knowledgeable individuals,, contact them immediately. The sooner you can investigate and "clean" your computer, the less likely it is to cause additional damage to other computers on the network. If you are on a home computer or laptop, disconnect your computer from the internet; this will prevent the attacker from accessing your system.
- **Remove the malicious code.** If you have antivirus software installed on your computer, update

Protecting Your Business from Malicious Software

the software and perform a manual scan of your entire system. If you do not have antivirus software, you should purchase it online or in a computer store as soon as possible. If the software cannot locate and remove the infection, you may need to reinstall your operating system, usually with a system restore disk. Note that reinstalling or restoring the operating system typically erases all of your files and any additional software that you have installed on your computer.

- After reinstalling the operating system and any other software, install all of the appropriate patches to fix known vulnerabilities.

One thing is for sure, threats to your computer will continue to evolve. Although you cannot eliminate every hazard, by using caution, installing and using antivirus software, and following other simple security practices, you can significantly reduce your risk and strengthen your protection against malicious code.

Now read the main points from chapter 3 overleaf.

Main points from chapter 3

PROTECTING YOUR BUSINESS FROM MALICIOUS SOFTWARE

- Malicious software (Malware) is one of the biggest cyber threats facing small to medium size business (and also individuals)

- It is essential to install and maintain antivirus software. Antivirus software recognizes malware and protects your computer against it.

- Use caution with links and attachments. Take appropriate precautions when using email and web browsers to reduce the risk of an infection.

- Be wary of unsolicited email attachments and use caution when clicking on email links, even if they seem to come from people you know.

- Install or enable a firewall. Firewalls can prevent some types of infection by blocking malicious traffic before it enters your computer. Some operating systems include a firewall; if the operating system you are using includes one, enable it.

- Use anti-spyware tools. Spyware is a common virus source, but you can minimize infections by using a program that identifies and removes spyware. Most antivirus software includes an anti-spyware option; ensure you enable it.

- Avoid using public Wi-Fi. Unsecured public Wi-Fi may allow an attacker to intercept your device's network traffic and gain access to your personal information.

- Remove the malicious code. If you have anti-virus software installed on your computer, update the software and perform a manual scan of your entire system.

- If you do not have antivirus software, you should purchase it online or in a computer store as soon as possible.

- If the software cannot locate and remove the infection, you may need to reinstall your operating system, usually with a system restore disk.

Part 4

PROTECTING EMAILS

Chapter 4

Cyber Security and Emails

One very important route for malicious code is through your emails. It is very important to be aware of the potential problems associated with emails.

Why can email attachments be dangerous?
Some characteristics that make email attachments convenient and popular also make them a common tool for attackers:

- Email is easily circulated – forwarding email is so simple that viruses can quickly infect many machines. Most viruses do not even require users to forward the email – they scan a users' mailbox for email addresses and automatically send the infected message to all of the addresses they find. Attackers take advantage of the reality that most users will automatically trust and open a message that comes from someone they know.

- Email programs try to address all users' needs–almost any type of file can be attached to an email message, so attackers have more freedom with the types of viruses they can send.

- Email programs offer many "user-friendly" features. Some email programs have the option to automatically download email attachments, which immediately exposes your computer to viruses within the attachments.

What steps can you take to protect yourself and others in your address book?

- **Be wary of unsolicited attachments, even from people you know.** Just because an email message looks like it came from someone you know does not mean that it did. Many viruses can "spoof" the return address, making it look like the message came from someone else. If you can, check with the person who supposedly sent the message to make sure it's legitimate before opening any attachments. This includes email messages that appear to be from your internet service provider (ISP) or software vendor

- **Keep software up to date.** Install software patches so that attackers can't take advantage of known problems or vulnerabilities. Many operating systems offer automatic updates. If this option is available, you should enable it.

- **Trust your instincts.** If an email or email attachment seems suspicious, don't open it, even if your antivirus software indicates that the message is clean. Attackers are constantly releasing new viruses, and the antivirus software might not have the signature. At the very least, contact the person who supposedly sent the message to make sure it's legitimate before you open the attachment. However, especially in the case of forwards emails, even messages sent by a legitimate sender might contain a virus. If something about the email or the attachment makes you uncomfortable, there may be a good reason. Don't let your curiosity put your computer at risk.

- **Save and scan any attachments before opening them.** If you have to open an attachment before you can verify the source, take the following steps:
  1. Be sure the signatures in your antivirus software are up to date.
  2. Save the file to your computer or a disk.
  3. Manually scan the file using your antivirus software.

4. If the file is clean and doesn't seem suspicious, go ahead and open it.

- **Turn off the option to automatically download attachments.** To simplify the process of reading email, many email programs offer the feature to automatically download attachments. Check your settings to see if your software offers the option, and make sure to disable it.

- **Consider creating separate accounts on your computer.** Most operating systems give you the option of creating multiple user accounts with different privileges. Consider reading your email on an account with restricted privileges. Some viruses need "administrator" privileges to infect a computer.

- **Apply additional security practices.** You may be able to filter certain types of attachments through your email software.

- Patches are software and operating system (OS) updates that address security vulnerabilities within a program or product. Software vendors may choose to release updates to fix performance bugs, as well as to provide enhanced security features.

Now read the main points from Chapter 4.

Main points from Chapter 4

CYBER SECURITY AND EMAILS

- One very important route for malicious code is through your emails. It is very important to be aware of the potential problems associated with emails.

- Some characteristics that make email attachments convenient and popular also make them a common tool for attackers:

- Email is easily circulated–forwarding email is so simple that viruses can quickly infect many machines.

- Most viruses do not even require users to forward the email – they scan a users' mailbox for email addresses and automatically send the infected message to all of the addresses they find.

- There are steps you can take to protect yourself and others in your address book. For example be wary of unsolicited attachments, even from people you know. Many viruses can "spoof" the return address, making it look like the message came from someone else.

- Keep software up to date. Install software patches so that attackers can't take advantage of known problems or vulnerabilities.

- Trust your instincts. If an email or email attachment seems suspicious, don't open it, even if your antivirus software indicates that the message is clean.

- Save and scan any attachments before opening them. If you have to open an attachment before you can verify

- Turn off the option to automatically download attachments.

- Consider creating separate accounts on your computer. Most operating systems give you the option of creating multiple user accounts with different privileges

- Apply additional security practices. You may be able to filter certain types of attachments through your email software

Part 5

MANAGING PASSWORDS

Chapter 5

The Use of Effective Passwords to Protect Data

Why you need strong passwords
You probably use personal identification numbers (PINs), passwords, or passphrases every day: from getting money from the ATM or using your debit card in a store, to logging in to your business computers, email or into an online retailer. Tracking all of the number, letter, and word combinations may be frustrating, but these protections are important because hackers represent a real threat to your information, either personal or business related. Often, an attack is not specifically about your account, but about using the access to your information to launch a larger attack. Password hacking is often carried out in one of the following ways:

- Your password can be cracked by cycling through every possibility until it matches your character combination. You can, however, make it take a very long time by using a complex password.

- **Dictionary.** With this method of hacking, a hacker will run a defined 'dictionary' against your passwords. This dictionary also includes the most common password combinations, therefore it is a relatively easy and quick way of hacking into weakly protected accounts.

- By using a single-use, strong password for each account, you should be able to protect yourself from a dictionary hack.

- **Phishing and social engineering (see chapter 7).** Accessing someone's password using a phishing or social engineering attack is not technically a type of hack, but it provides the 'hacker' with access to your passwords and confidential information. This in turn allows them to access your accounts. Phishing occurs when a hacker targets you with spoofed *Brute force attacks.*

- A hacker uses automated software to guess your username and password combination.

- The software tries character combinations and will try the most commonly used passwords first, so weak or common passwords can be relatively simple for a brute force attack to crack.

The repercussions of identity theft can be long lasting and they are not only limited to financial problems. The victim could also face a range of emotional implications, including stress and anxiety. Therefore, it's important that you take measures to protect yourself from the burdens of having an account hacked.

One of the best ways to protect information or physical property is to ensure that only authorized people have access to it. Verifying that those requesting access are the people they claim to be is the next step. This authentication process is more important and more difficult in the cyber world. There are simple guidelines on Microsoft and Google to enable you to do this. Passwords are the most common means of authentication, but only work if they are complex and confidential. Many systems and services have been successfully breached because of non-secure and inadequate passwords. Once a system is compromised, it is open to exploitation by other unwanted sources.

How to choose good passwords
Most people use passwords that are based on personal information and are easy to remember. However, that also makes it easier for an attacker to crack them.

Consider a four-digit PIN. Is yours a combination of the month, day, or year of your birthday? Does it contain your address or phone number? Think about how easy it is to find someone's birthday or similar information. What about your email password – is it a word that can be found in the dictionary? If so, it may be susceptible to dictionary attacks, which attempt to guess passwords based on common words or phrases.

Although intentionally misspelling a word ("daytt" instead of "date") may offer some protection against dictionary attacks, an even better method is to rely on a series of words and use memory techniques, or mnemonics, to help you remember how to decode it. For example, instead of the password "hoops," use "IlTpbb" for "[I] [l]ike [T]o [p]lay [b]asket[b]all." Using both lowercase and capital letters adds another layer of obscurity. Changing the same example used above to "Il!2pBb." creates a password very different from any dictionary word.

Length and complexity
You should consider using the longest password or passphrase permissible (8–64 characters) when you can.

For example,"Pattern2baseball#4mYmiemale!" would be a strong password because it has 28 characters and includes the upper and lowercase letters, numbers, and special characters. You may need to try different variations of a passphrase – for example, some applications limit the length of passwords and some do not accept spaces or certain special characters. Avoid common phrases, famous quotations, and song lyrics.

Do's and don'ts

Once you've come up with a strong, memorable password it's tempting to reuse it-this is not advised! Reusing a password, even a strong one, endangers your accounts just as much as using a weak password. If attackers guess your password, they would have access to your other accounts with the same password. Use the following techniques to develop unique passwords for each of your accounts:

- Use different passwords on different systems and accounts.
- Use the longest password or passphrase permissible by each password system.
- Develop mnemonics to remember complex passwords.
- Consider using a password manager program to keep track of your passwords. (See below.)
- Do not use passwords that are based on personal information that can be easily accessed or guessed.

- Do not use words that can be found in any dictionary of any language.

How to protect your passwords
Help your staff cope with password overload
If you're in charge of how passwords are used in your organisation, there are a number of things you can do that will improve security. Most importantly, your staff will have dozens of non-work related passwords to remember as well, so only enforce password access to a service if you really need to.

Where you do use passwords to access a service, do not enforce regular password changes. Passwords really only need to be changed when you suspect a compromise of the login credentials.

After choosing a password that's easy to remember but difficult for others to guess, do not write it down and leave it someplace where others can find it. Writing it down and leaving it in your desk, next to your computer, or, worse, taped to your computer, makes it easily accessible for someone with physical access to your office. Do not tell anyone your passwords, and watch for attackers trying to trick you through phone calls or email messages requesting that you reveal your passwords.

Password managers

Programs called password managers offer the option to create randomly generated passwords for all of your accounts. You then access those strong passwords with a master password. If you use a password manager, remember to use a strong master password.

Password problems can stem from your web browsers' ability to save passwords and your online sessions in memory. Depending on your web browsers' settings, anyone with access to your computer may be able to discover all of your passwords and gain access to your information. Always remember to log out when you are using a public computer (at the library, an internet cafe, or even a shared computer at your office). Avoid using public computers and public Wi-Fi to access sensitive accounts such as banking and email.

There's no guarantee that these techniques will prevent an attacker from learning your password, but they will make it more difficult.

Don't forget security basics

- Keep your operating system, browser, and other software up to date.
- Use and maintain antivirus software and a firewall.

- Regularly scan your computer for spyware. (Some antivirus programs incorporate spyware detection.)

- Use caution with email attachments and untrusted links.

- Watch for suspicious activity on your accounts.

Now read the main points from Chapter 5.

Main points from Chapter 5

PROTECTING PASSWORDS

- It is very important indeed for your business to have a strong and clear password policy. The hacking of passwords is very common and can result in the loss of key data.

- Often, an attack is not specifically about your account, but about using the access to your information to launch a larger attack.

- Your password can be cracked by cycling through every possibility until it matches your character combination. You can make it take a very long time by using a complex password.

- A hacker will run a defined 'dictionary' against your passwords. This dictionary also includes the most common password combinations, therefore it is a relatively easy and quick way of hacking into weakly protected accounts.

- By using a single-use, strong password for each account, you should be able to protect yourself from a dictionary hack.

- Phishing and social engineering. Accessing someone's password using a phishing or social engineering attack is not technically a type of hack, but it provides the 'hacker' with access to your passwords and confidential information. This in turn allows them to access your accounts. Phishing occurs when a hacker targets you with spoofed Brute Force attacks.

- If you're in charge of how passwords are used in your organisation, there are a number of things you can do that will improve security. Most importantly, your staff will have dozens of non-work related passwords to remember as well, so only enforce password access to a service if you really need to.

- Programs called password managers offer the option to create randomly generated passwords for all of your accounts. You then access those strong passwords with a master password. If you use a password manager, remember to use a strong master password.

Part 6

UTILISING FIREWALLS

Chapter 6

Using Firewalls to Protect Your System

Understanding Firewalls

When your computer is accessible through an internet connection or Wi-Fi network, it is susceptible to attack. However, you can restrict outside access to your computer – and the information on it – with a firewall.

What do firewalls do?

Firewalls provide protection against cyber attackers by shielding your computer or network from malicious or unnecessary network traffic. Firewalls can also prevent malicious software from accessing a computer or network via the internet. Firewalls can be configured to block data from certain locations (i.e., computer network addresses), applications, or ports while allowing relevant and necessary data through.

There are a number of firewalls to choose from and having an understanding of each type is important.

What type of firewall is best?
Categories of firewalls include hardware and software. While both have advantages and disadvantages, the decision to use a firewall is more important than deciding which type you use.

- **Hardware** – Typically called network firewalls, these physical devices are positioned between your computer and the internet (or other network connection). Many vendors and some internet service providers (ISPs) offer integrated small office / home office routers that also include firewall features. Hardware-based firewalls are particularly useful for protecting multiple computers and controlling the network activity that attempts to pass through them.

- The advantage of hardware-based firewalls is that they provide an additional line of defense against attacks reaching desktop computing systems. The disadvantage is that they are separate devices that require trained professionals to support their configuration and maintenance.

- **Software** – Most operating systems (OSs) include a built-in firewall feature that you should enable for added protection, even if you have an external firewall. Firewall software is also available separately from your local computer store, software vendor, or ISP. If you download firewall software from the internet, make sure it is from a reputable source (i.e., an established software vendor or service provider) and offered via a secure site. The advantage of software firewalls is their ability to control the specific network behavior of individual applications on a system. A significant disadvantage of a software firewall is that it is typically located on the same system that is being protected. Being located on the same system can hinder the firewall's ability to detect and stop malicious activity. Another possible disadvantage of software firewalls is that – if you have a firewall for each computer on a network – you will need to update and manage each computer's firewall individually.

*

What configuration settings to apply?
Most commercially available firewall products, both hardware and software based, come pre-configured and ready to use. Since each firewall is different, you will need to read and understand the documentation that comes with it to determine whether the default firewall settings are sufficient for your needs. This

is particularly concerning because the "default" configuration is typically less restrictive, which could make your firewall more susceptible to compromise. Alerts about current malicious activity sometimes include information about restrictions you can implement through your firewall.

Though properly configured firewalls may effectively block some attacks, do not be lulled into a false sense of security. Firewalls do not guarantee that your computer will not be attacked. Firewalls primarily help protect against malicious traffic, not against malicious programs (i.e., malware), and may not protect you if you accidentally install or run malware on your computer. However, using a firewall in conjunction with other protective measures (e.g., antivirus software and safe computing practices) will strengthen your resistance to attacks.

Below are listed the best types of firewalls for businesses

- Windows Defender or OS X Application Firewall- this is best for Individual entrepreneurs

- Third-party software firewall-this is best for individuals handling sensitive data

- Firewall & antivirus software-This is best for small businesses

Using Firewalls to Protect Your System

- Basic router- This is the best budget option

- Firewall router-This is best for medium-sized businesses

- VPN router-This is best for businesses with multiple locations

- Load balancer-This is best for businesses hosting their own websites

- Unified threat management (UTM) This is best for large businesses

The above options are described below.

Windows Defender or OS X Application Firewall. Most people don't realize that their Windows or Mac computer already includes free firewall software. So if you're an individual running a small business on your own, you may already have all the intrusion protection you need – no expensive third-party firewall necessary.

If you have a Windows computer, your operating system already includes Windows Defender – Microsoft's free firewall software. Windows Defender is a stateful inspection firewall, so it analyzes both the TCP handshake and packet labels (more on those later) on every online exchange. It comes pre-enabled on your computer, so you don't have to do anything to get started.

On Apple computers, you get the OS X Application Firewall – a circuit level gateway software that monitors TCP handshakes. While it allows you to set your own firewall rules, it doesn't use packet filtering, which makes it a bit less reliable than the free Windows firewall. And it isn't pre-enabled, so be sure to turn your firewall software on before connecting to the internet.

Keep in mind, too, that both Windows Defender and the OS X Application Firewall are *software firewalls*, so they can protect only your individual computer – hence the reason they are recommended for individuals, not for larger companies. They're also fairly basic, so if you're handling a lot of sensitive data (like customer credit card numbers, addresses, or phone numbers), you may want to upgrade to a third-party software firewall.

Third-party software firewall
Third-party firewalls complement the existing firewall software on your computer. They deliver extra security features to help thwart would-be cyber criminals. Every third-party firewall solution offers a different combination of features, so you may have to do some shopping to find the right software for your needs. But features can include an extra layer of deep packet inspection, anti-spam functions, data backup, and more.

This option is recommended if you're an individual handling sensitive data because it gives you additional tools and protections to keep that data safe – while still being affordable and manageable. That being said, companies with multiple employees may prefer a hardware firewall. Since a software firewall can protect only the devices the software is installed on, it doesn't protect your entire network. Plus, you have to manually install and update the software on each device on your network (even mobile devices).

Depending on the software, you may also have to buy separate licenses for each device, which gets pricey – especially considering how expensive many third-party firewalls are.

Having a software firewall on your company devices is still important. If each device on your network has a software firewall, your network's still protected if one device is infiltrated. Software firewalls also allow your employees to work from home or elsewhere and enjoy the same online security they get in the office.

Firewall + antivirus software: Best for small businesses
The more employees you have, the more likely it is that someone on your network will accidentally install malware or download a computer virus. That's why the best business firewall is a firewall + antivirus software combination. Firewalls that include antivirus software use deep packet inspection to identify and reject files, messages, and other forms of data that include malware

or viruses. Consequently, they have a better record of intrusion detection than a regular firewall. And usually, this kind of software acts as a web application firewall, so it keeps you safe no matter which app you use to access the internet.

Basic router: Best budget option

If you're running a small business with multiple employees, chances are you've already invested in a basic Wi-Fi router so everyone in the office can connect to the internet at once. If so, you've already got basic firewall protection. A Wi-Fi router is a great low-budget, small-business firewall solution because routers automatically block any external traffic that doesn't meet basic security parameters (set by you, of course). That essentially makes your router a stateless firewall, monitoring TCP handshakes, (short for *transmission control protocol)* to make sure every incoming request is on "the list" for your internal network.

Of course, that means your router offers only minimal network security – hardly ideal if you're handling a lot of sensitive data you don't want compromised. In that case, you should probably upgrade to a firewall router or a third-party software firewall.

On the upside, though, a router is a hardware firewall, so it protects all the devices on your network. That saves you money, since you don't have to buy licenses for each employee's computer.

Firewall router: Best for medium-sized businesses
As your business grows, installing and maintaining firewall software on each employee device becomes more and more impractical – at least as your primary form of network security. In that case, a hardware firewall that protects your whole network at once may be a better solution. Firewall routers upgrade the security you get with a basic router by adding more complex firewall rules to better identify security threats. Some models offer stateful security firewalls, built-in antivirus software (which operates from your router, not individual devices), application monitoring, and "parental" controls to block employees from accessing dangerous sites (or anything you deem to be inappropriate for work). All that means you get all the protection of a software firewall, but you can control all your settings and updates in one device. Plus, you get protection for every device connected to your Wi-Fi network – including mobile devices.

VPN Best for businesses with multiple locations
If your business is spread across multiple offices or you have remote employees, you know how difficult it can be

to keep everybody on the same page. However, with a VPN router, it gets a lot easier and boosts your security.

Normally, your internal network is accessible only to devices on your internet connection. That means the devices have to be physically present in the same location to connect to each other for file sharing, printing, and other internal network functions.

But with a virtual private network you can extend your private, internal business network to other approved devices and networks via VPN tunnels. These tunnels act as another layer of data layering (like putting a letter inside an envelope inside a box), filtering out attacks from hackers trying to infiltrate your internal network connection. VPN routers simplify the process. When each of your locations uses a VPN router, your routers can communicate with each other, effectively combining the internal networks of each office into one big private network. In the end, that makes it easier to communicate and collaborate with your remote employees and offices while still enjoying a high level of cybersecurity company-wide.

Load balancer: Best for businesses hosting their own websites

If your business hosts websites on your own servers, you'll probably need a load balancer in addition to your private network firewall solution.

When hosting websites, your servers need to be external-facing – meaning the public can access the data stored on your servers. Otherwise, users won't be able to load your websites. But you also want to protect your servers from hackers and other malicious online entities. Fortunately, a load balancer can act as an automatic firewall, much like a router for your internal network.

Load balancers distribute incoming traffic across your servers. That way, no single server gets overwhelmed with simultaneous requests. Not only does load balancing make your hosted websites load faster – it can also protect your business from DDOS attacks (where hackers hijack multiple systems to overwhelm your server and crash your site).

If you're already using load balancing, you may not need another firewall to protect your servers. Load balancers already monitor TCP handshakes and perform packet filtering functions to determine the most efficient way to distribute incoming requests. In other words, it already acts as a stateful firewall and discards malicious incoming traffic.

Unified threat management (UTM): Best for large businesses

If you run a large business, chances are you need a more complex security solution than a router or single software. In that case, you may want to consider a unified threat management (UTM) solution.

Each UTM product is different – some are physical devices, some are software, some are cloud-based, and some are a combination of all three. Whatever the implementation method, though, all UTM solutions aim to offer a one-stop shop for all your security needs.

UTM solutions usually offer firewall, antivirus, VPN, and other intrusion detection and prevention features in one place. That way, you get deep packet filtering for *all* web applications on all the devices on your Wi-Fi network (or your virtual private network), but it's all controlled in one place.

The exact cost of a UTM solution can vary dramatically depending on the provider you choose, the size of your business, and the specific combination of features your UTM includes. Some UTMs cost roughly the same as a third-party software firewall. But be prepared for higher overall costs – you are rolling antivirus protection, VPN security, software firewalls, and hardware firewalls together into one solution, after all.

One such company offering this kind of solution is Sophos, a very well known company in the cyber-security world.
https://www.sophos.com/Security/Firewall

Firewall terms to know

Firewall providers use a lot of jargon, which makes it hard to understand what each option actually offers. Below is a brief breakdown of some of the terms used a lot in this chapter (some of which may be repeated in the glossary of terms at the back of the book).

TCP handshake

TCP may sound like a drug or a high-end cleaning product, but it's actually short for *transmission control protocol*. Every online device uses TCP to connect to the internet, and when two devices want to connect to each other, they use a TCP handshake.

Clever hackers can fake a TCP handshake and use it to get access to your business's internal network. That's one of the most elementary reasons why firewall protection is so important.

Circuit-level gateways

Some firewalls act as circuit-level gateways, which means they monitor TCP handshakes on your device or network to determine whether those sessions are legitimate or not. This type of web filtering is a pretty basic security solution, but it can help protect you from hackers who attempt to fake a TCP handshake to gain access to your company's private network.

Circuit-level gateways also mask the individual IP

addresses of each device on your network. Instead, all outgoing traffic from your network is given an ID that goes with the IP address for your circuit-level gateway device (usually a router).

This provides an extra level of privacy for your company and your employees.

Packet filtering

Data on the internet is transmitted via *packets*. Think of a packet like an envelope: the outside is labeled with delivery information (the delivery address, return address, etc.), while the inside contains the actual message. Once a TCP handshake is complete, the website you're trying to access sends a data packet. The packet is labeled with your IP address (the delivery address) and the source IP (the sender's address), and it contains a small amount of data that your computer uses to load the page. Packet filtering is a security process in which your firewall examines the labels on the outside of any data packets being sent to your IP address. Packet filtering security solutions use a predefined set of firewall rules (controlled by you) to determine (based on the packet labels) whether incoming traffic is malicious or not. If it's malicious, the firewall discards the packet – thereby denying access to your network and protecting you from hackers.

Stateless firewall

A *stateless firewall* is a firewall that uses only packet filtering to monitor your online connections. While packet filtering is certainly an effective method of blocking malicious traffic to your network, it's still fairly basic since it takes only the outside labels of incoming data packets into account.

So a stateless firewall – while generally effective – doesn't use more complex encryption to identify fraudulent connections. That said, stateless firewalls may still come with other security features (like application monitoring), so you certainly shouldn't rule them out.

Stateful inspection firewall

A stateful inspection firewall is a bit more complex because it combines the TCP handshake monitoring and packet label inspection of basic packet filtering. That makes a stateful inspection firewall more secure than a circuit-level gateway or stateless firewall alone, but it does require extra computing resources.

So if your small business can't afford the newest devices, a stateful firewall may slow down your computers and internet loading speeds.

Proxy deep packet inspection

A security firewall that uses deep packet inspection opens data packets on your behalf (by proxy) so it can analyze the actual contents inside and identify viruses, malware, or other threats. That means it would protect you and your employees anytime someone in your network tries to access a shady application or click on a malicious link in an email.

The takeaway

In the end, the best type of firewall for your business depends on your needs. If you're running a small business (or you're the sole employee in your business) and don't handle a lot of sensitive data, a basic solution (like a Wi-Fi router or the free firewall software included with your computer) may be the easiest and most cost-efficient solution.

If you are running a small to medium-sized business or have heightened security needs, you may want to invest in a third-party software firewall, an antivirus + firewall combination, or a firewall router. If working with employees in multiple locations, a VPN router may be the way to go.

Larger businesses, though, may need the more intense firewall protection included with a unified threat management (UTM) solution. And companies that host

Using Firewalls to Protect Your System

websites will definitely want to protect their server with a load balancer (in addition to their internal network firewall).

Now read the main points from chapter 6 overleaf.

Main points from Chapter 6

USING FIREWALLS

- When your computer is accessible through an internet connection or Wi-Fi network, it is susceptible to attack. However, you can restrict outside access to your computer – and the information on it – with a firewall.

- Firewalls provide protection against outside cyber attackers by shielding your computer or network from malicious or unnecessary network traffic.

- Firewalls can also prevent malicious software from accessing a computer or network via the internet.

- Firewalls can be configured to block data from certain locations (i.e., computer network addresses), applications, or ports while allowing relevant and necessary data through.

- Categories of firewalls include hardware and software. While both have advantages and disadvantages, the decision to use a firewall is more important than deciding which type you use.

Part 7

PHISHING AND OTHER ATTACKS

Chapter 7

Avoiding Phishing and Other Attacks

In this chapter, we discuss the dangers of the various types of attacks, such as social engineering, phishing, vishing and smishing to your business.

What is a social engineering attack?
In a social engineering attack, an attacker uses human interaction (social skills) to obtain or compromise information about an organization or its computer systems. An attacker may seem unassuming and respectable, possibly claiming to be a new employee, repair person, or researcher and even offering credentials to support that identity. However, by asking questions, he or she may be able to piece together enough information to infiltrate an organization's network. If an attacker is not able to gather enough information from one source, he or she may contact another source within

the same organization and rely on the information from the first source to add to his or her credibility.

What is a phishing attack?
Phishing is a form of social engineering. Phishing attacks use email or malicious websites to solicit personal information by posing as a trustworthy organization. For example, an attacker may send email seemingly from a reputable credit card company or financial institution that requests account information, often suggesting that there is a problem. When users respond with the requested information, attackers can use it to gain access to the accounts. Phishing attacks may also appear to come from other types of organizations, such as charities. Attackers often take advantage of current events and certain times of the year, such as

- Natural disasters
- Economic concerns
- Major political elections
- Holidays

What is a vishing attack?
Vishing is the social engineering approach that leverages voice communication. This technique can be combined with other forms of social engineering that entice a victim to call a certain number and divulge sensitive

information. Advanced vishing attacks can take place completely over voice communications by exploiting Voice over Internet Protocol (VoIP) solutions and broadcasting services. VoIP easily allows caller identity (ID) to be spoofed, which can take advantage of the public's misplaced trust in the security of phone services, especially landline services.

Landline communication cannot be intercepted without physical access to the line; however, this trait is not beneficial when communicating directly with a malicious actor.

What is a smishing attack?

Smishing is a form of social engineering that exploits SMS, or text, messages. Text messages can contain links to such things as webpages, email addresses or phone numbers that when clicked may automatically open a browser window or email message or dial a number. This integration of email, voice, text message, and web browser functionality increases the likelihood that users will fall victim to engineered malicious activity.

Common indicators of phishing attempts

- **Suspicious sender's address.** The sender's address may imitate a legitimate business. Cybercriminals often use an email address that closely resembles one from a reputable company by altering or omitting a few characters.

- **Generic greetings and signature.** Both a generic greeting – such as "Dear Valued Customer" or "Sir/Madam" – and a lack of contact information in the signature block are strong indicators of a phishing email. A trusted organization will normally address you by name and provide their contact information.

- **Spoofed hyperlinks and websites.** If you hover your cursor over any links in the body of the email, and the links do not match the text that appears when hovering over them, the link may be spoofed. Malicious websites may look identical to a legitimate site, but the URL may use a variation in spelling or a different domain (e.g., .com vs. .net). Additionally, cybercriminals may use a URL shortening service to hide the true destination of the link.

- **Spelling and layout.** Poor grammar and sentence structure, misspellings, and inconsistent formatting are other indicators of a possible phishing attempt. Reputable larger institutions have dedicated personnel that produce, verify, and proofread customer correspondence. Most smaller companies take great care when sending out correspondence.

- **Suspicious attachments.** An unsolicited email requesting a user download and open an attachment is a common delivery mechanism for malware.

- A cybercriminal may use a false sense of urgency or importance to help persuade a user to download or open an attachment without examining it first.

How do you avoid being a victim?

- Be suspicious of unsolicited phone calls, visits, or email messages from individuals asking about employees or other internal information. If an unknown individual claims to be from a legitimate organization, try to verify his or her identity directly with the company.

- Do not provide personal information or information about your organization, including its structure or networks, unless you are certain of a person's authority to have the information.

- Do not reveal personal or financial information in email, and do not respond to email solicitations for this information. This includes following links sent in email. Don't send sensitive information over the internet before checking a website's security.

- Pay attention to the Uniform Resource Locator (URL) of a website. Look for URLs that begin with "https" – an indication that sites are secure – rather than "http." Look for a closed padlock icon – a sign your information will be encrypted.

- If you are unsure whether an email request is legitimate, try to verify it by contacting the company directly. Do not use contact information provided on a website connected to the request, instead, check previous statements for contact information.

- Install and maintain anti-virus software, firewalls, and email filters to reduce some of this traffic.

- Take advantage of any anti-phishing features offered by your email client and web browser.

- Enforce multi-factor authentication (MFA).

What to do if you think you are a victim?

- If you believe you might have revealed sensitive information about your organization, report it to the appropriate people within the organization, including network administrators. They can be alert for any suspicious or unusual activity.

Avoiding Phishing and Other Attacks

- If you believe your financial accounts may be compromised, contact your financial institution immediately and close any accounts that may have been compromised. Watch for any unexplainable charges to your account.
- Immediately change any passwords you might have revealed. If you used the same password for multiple resources, make sure to change it for each account, and do not use that password in the future.
- Watch for other signs of identity theft
- Consider reporting the attack to the police

Now read the main points from Chapter 7 overleaf.

Main points from Chapter 7

AVOIDING PHISHING AND OTHER ATTACKS

- There are dangers of various types of attacks, such as social engineering, phishing, vishing and smishing to your business. It is crucial to be aware of these and how to combat them.

- In a social engineering attack, an attacker uses human interaction (social skills) to obtain or compromise information about an organization or its computer systems.

- Phishing is a form of social engineering. Phishing attacks use email or malicious websites to solicit personal information by posing as a trustworthy organization.

- Vishing is the social engineering approach that leverages voice communication. This technique can be combined with other forms of social engineering that entice a victim to call a certain number and divulge sensitive information.

- Smishing is a form of social engineering that exploits SMS, or text, messages. Text messages can contain links to such things as webpages, email addresses or phone numbers that when clicked may automatically open a browser window or email message or dial a number.

- To avoid being a victim, be suspicious of unsolicited phone calls, visits, or email messages from individuals asking about employees or other internal information.

- If an unknown individual claims to be from a legitimate organization, try to verify his or her identity directly with the company.

- Do not provide personal information or information about your organization, including its structure or networks, unless you are certain of a person's authority to have the information.

- Do not reveal personal or financial information in email, and do not respond to email solicitations for this information.

Part 8

LOOKING AFTER MOBILES, TABLETS AND OTHER DEVICES

Chapter 8

Protecting Mobiles, Tablets and Other Portable Devices

Keep your Company Devices Safe from Threats
Gone are the days when the most sensitive information on an employee's phone was contact names and phone numbers. Now a smartphone or tablet can be used to gain access to anything from emails to stored passwords to proprietary company data and trade secrets. With the advent of 5G technology making accessibility easier and faster, more and more companies are poised to adopt mobile technology as a normal part of business.

Depending on how your organization uses such devices, unauthorized access to a smartphone, tablet or other device can lead to a catastrophic cyber-incident involving an organizations entire IT infrastructure. While it's important to implement cybersecurity safeguards as a whole, the following measures will help you avoid security issues with mobile devices in particular and keep your data safe.

1. Establish a Mobile Device Security Policy

Before issuing smartphones or tablets to your employees, establish a device usage policy. Provide clear rules about what constitutes acceptable use. Include what actions will follow if employees violate the policy. It is important that employees understand the security risks of smartphone use and the security measures they can take to mitigate those risks. Well-informed, responsible users are your first line of defence against cyber-attacks.

2. Establish a Bring Your Own Device Policy

If you allow employees to use their personal devices for company business, make sure you have a formal Bring Your Own Device (BYOD) policy in place. Your BYOD security plan should include:

- Requirements for installing remote wiping software on any personal devices used to store or access company data

- Education and training for employees on how to safeguard company data when they access wireless networks from their own mobile phones and devices

- Data protection practices that include requiring strong passwords and automatic locking after periods of inactivity

- Protocols for reporting lost or stolen devices
- The use of certain antivirus and protective security software
- Requirements for regular backups
- An approved list for those who want to download apps

3. Keep the Devices Updated with Current Software and Antivirus Programs

Software updates to mobile devices often include patches for various security holes that can be an open door for mobile malware and other security threats. Therefore, it is a security best practice to install the updates as soon as they become available. When it comes to antivirus software for mobile devices, there are many options to choose from and it can come down to preference. Some are free to use from the app store while others charge a monthly or annual fee and often come with better support. In addition to antivirus support, many of these programs will monitor Short Message Service (SMS) texts, Multimedia Messaging Service (MMS) and call logs for suspicious activity. They can use blacklists to prevent users from installing known malware to their devices.

4. Backup Device Content on a Regular Basis

Just as you backup your computer data regularly, so should you backup data on your company's mobile devices. If a device is lost or stolen, you'll have peace of mind knowing your valuable data is safe and that it can be restored.

5. Choose Passwords Carefully

As we discussed in the chapter on passwords, in the UK, the average email address is associated with numerous online accounts yet the average internet user reuses a handful of passwords to protect them all. Obviously, this lack of security awareness is what hackers count on to steal data.

It is recommended that businesses use the following tips to ensure mobile device passwords are easy to remember and hard to guess.

- Require employees to change the device's login password at least every 90 days

- Implement two-factor authentication to verify a user's identity

- Passwords should be at least eight characters long and include uppercase and lowercase letters, numbers and special characters, such as asterisks, exclamation points and pound signs

- Don't use simple number sequences such as "12345," or names of spouses, children or pets in a password. A hacker can spend just a couple minutes on a social media site to figure out this information.

- Because of the convenience they offer, smartphones and tablet devices have become a constant presence in the modern business world. As usage soars, it becomes increasingly important to take steps to protect your company and its sensitive data from mobile threats, both new and old.

- Lastly, even with the best security solutions in place, there's never a 100% guarantee. It's important to protect your company from the liability risks associated with a cyber-attack, whether it be through an employee's mobile device or your company server.

Apple iOS and Google Android have both improved their security over the years, but you can also download a free or paid mobile antivirus app to increase your protection.

Mobile security software

You don't always have to pay to get effective extra protection for mobile devices. Free apps, such as those available from companies like Avast, AVG, and McAfee, are popular for obvious reasons. Some free apps do all the basics well. If your only concerns are malware, a free app is the obvious choice. However, free antivirus downloads won't come with the same functions as their paid-for cousins. Plus, free apps sometimes only have full functionality for a few weeks, and then you're left with only the basic tools and features after the trial period ends. If you have expensive Android devices, it may be worth investing in a paid-for security app. As well as offering malware protection and anti-spam tools, they often offer useful features, such as cloud-based (online) backup and the ability to remotely wipe a lost or stolen device.

Mobile security software features
While free mobile security apps will usually only offer very basic functions, if you pay for software you can access a range of more enhanced premium features. Although each package will vary, below are some typical features on offer.

- Anti-phishing: Sniffs out dodgy-looking links and should bar you from accessing potentially dangerous websites designed to infect machines or steal your personal data.

- App lock: Lets you protect certain apps and settings with a password to restrict who can use or edit them.

- Backup: Preserves your personal information by saving it online using 'cloud' (internet-based) storage, as well as on your device. Backup happens either according to a schedule or before you complete a remote wipe. Backups can then be restored to any compatible device.

- Call/SMS blocking: Filters and blocks unwanted calls and text messages, with warnings when you receive a suspicious text. Parental controls: Lets you prevent access to certain types of content.

- Privacy adviser: Scans and checks any apps you try to download, telling you which ones are asking for more access than they really need.

- Remote location: Uses GPS to show your phone or tablet's location on a map from any web browser.

- Remote lock: Locks down your device remotely via SMS or web browser interface, preventing unauthorised access. Some apps even allow you to create a customised lock-screen message that displays your contact details to aid the safe return of your device.

- Remote photo: Helps identify unauthorised users of your phone or tablet by taking their photo and sending it to you via email. Some apps take snaps through the device's camera and emails them to you when a password is entered incorrectly a few times.

- Remote wipe: Preserves your privacy by enabling you to wipe all contacts, calendars, photos, texts, browsing history and memory card in your phone.

- Safe browsing mode: Checks webpages in real time, as you're browsing the web, keeping you protected when you click on links or every time you type in a web address.

- SIM lock: Stops others from using your phone or tablet by locking it down when the SIM card is removed.

- Tune up: Optimises battery, data and storage use by identifying apps that use too much battery life or slow down your device's performance. This means you can shut down and uninstall those apps.

- Uninstall protection: Prevents thieves from bypassing the mobile security app and wiping

Protecting Mobiles, Tablets and Other Portable Devices

any device data and apps. A password is required to delete the app.

To summarise, as we have discussed mobile technology is being used more and more in the workplace therefore it is vitally important to ensure that you have adequate protection in place and have a very clear policy.

Now read the Main Points from Chapter 8 overleaf.

Main Points from Chapter 8

PROTECTING MOBILES, TABLETS AND OTHER PORTABLE DEVICES

- A smartphone or tablet can be used to gain access to anything from emails to stored passwords to proprietary company data and trade secrets.

- Depending on how your organization uses such devices, unauthorized access to a smartphone, tablet or other device can lead to a catastrophic cyber-incident involving an organizations entire IT infrastructure.

- You should establish a Mobile Device Security Policy. Also, Establish a Bring Your Own Device Policy. If you allow employees to use their personal devices for company business, make sure you have a formal Bring Your Own Device (BYOD) policy in place.

- Keep the Devices Updated with Current Soft-ware and Antivirus Programs

- Backup Device Content on a Regular Basis
- Choose Passwords Carefully

Conclusion

From the outset, this book has stressed the enormous importance of cyber security, both to small and medium size businesses and indeed all involved in handling data on whatever device, whether it is on an individual computer if you are a small business or solo entrepreneur, or whether it is on a network if you are a larger organization.

Cyber-attacks are becoming ever more prevalent and business must keep pace with the threat. In chapter one, we gave a basic overview of the meaning of cyber-security. In chapter two, we discussed the absolute importance of backing up. This might seem basic advice but it is amazing just how many businesses fail to do this, particularly smaller outfits.

In chapter three, we discussed the prevalence of Malware (Malicious software) and how to combat this. Chapter four covered the potential problems with emails and how to combat these problems, what to watch out for. Chapter five covered the potential issues with passwords and how to prevent hacking, which can expose a business (and individuals) to enormous problems.

Chapter six dealt with the importance of having effective firewalls and chapters seven and eight dealt with phishing and the importance of protecting mobile devices.

Above all, this book is what is known as a 'bread and butter' book. We have been conscious of the need to avoid blinding the readers with science, to avoid the use of terminology (as far as we can) and concentrate on explaining the different areas of risk and providing solutions. At the best, we would hope that this brief book can provide the basis for an effective cyber-security policy on which businesses can base their actions.

Good luck with your business and its ongoing security!

* * * *

Useful Websites

Useful websites listing the top cybersecurity companies for small to medium enterprises in the UK

https://www.networklondon.co.uk/top-cyber-security-companies

This website is extremely useful for identifying those companies which can tailor to your cyber-security needs as a small to medium size enterprise..

https://www.cybertango.io/cybersecurity-vendors/ Cybersecurity-UK

This site lists all of the most popular companies dealing with cybersecurity in the United Kingdom

*https://www.goodfirms.co/it-services/*cyber-security/uk

This site reviews the best cyber security companies in the UK in 2022.

https://www.ncsc.gov.uk/collection/small-business-guide

This is the website of the National Cyber Security Centre of the UK and gives comprehensive advice concerning business security

https://www.aon.com/Cyber-Solutions

The website of AON.Com a large cyber security company providing services to all areas of business

Glossary of Cyber Security Terms

Access control
Controlling who has access to a computer or online

Asset
Something of value to a person, business or organization.

Authentication
The process to verify that someone is who they claim to be when they try to access a computer or online service.

Backing up
To make a copy of data stored on a computer or server to lessen the potential impact at failure or loss

Bring your own device (BYOD)
The authorized use of personally owned mobile devices such as smartphone or tablets in the workplace.

Business continuity management
Preparing for and maintaining continued business operations following disruption or crisis.

Certification
Declaration that specified requirements have been met.

Certification body
An independent organization that provides certification services.

Cloud computing
Delivery of storage or computing services from remote services online (i.e. via the internet).

Data server
A computer or program that provides other computers with access to shared files over a network.

DMZ
Segment of a network where services accessed by less trusted users are isolated. The name is derived from the term "demilitarized zone".

Encryption
The transformation of data to hide its information content.

Ethernet
Communications architecture for wired local area networks based upon IEEE 802.3 standards.

Firewall
Hardware or software designed to prevent unauthorized access to a computer or network from another computer or network.

Hacker
Someone who violates computer security for malicious reasons, kudos or personal gain.

Intrusion detection system (IDS)
– Program or device used to detect that an attacker is or has attempted unauthorized access to computer resources. Intrusion prevention system (IPS) – Intrusion detection system that also blocks unauthorized access when detected.

Keyboard logger
A virus or physical device that logs keystrkes to secretly capture private information such as passwords or credit card details.

Local area network (LAN)
Communication network linking multiple computers within a defined location such as an office building.

Macro virus
Malware Malicious software that uses the macro capabilities of common applications such as spreadsheets and word processors to infect data.

Malware
Software intended to infiltrate and damage or disable computers.

Management system
A set of processes used by an organization to meet policies and objectives for that organization.

Network firewall
Device that controls traffic to and from a network.

Password
A secret series of characters used to authenticate a person's identity.

Personal firewall
Software running on a PC that controls network traffic to and from that computer.

Phishing
Method used by criminals to try to obtain financial or other confidential information (including user names and passwords) from internet users.

Portable device
A small easily transportable computing device such as a smartphone, laptop or tablet computer.

Proxy server
Server that acts as an intermediary between user and others servers, validating user requests.

Restore
The recovery of data following computer failure or loss.

Risk assessment
The process of identifying, analysing and evaluating risk.

Router
Device that directs messages within or between networks.

Screen scraper
A virus or physical device that logs information sent to a visual display to capture private or personal information.

Security control
Something that modifies or reduces one or more security risks.

Security information and event management (SIEM)
Process in which network information is aggregated, sorted and correlated to deter suspicious activities.

Security perimeter
A well-defined boundary within which security controls are enforced.

Server
Computer that provides data or services to other computers over a network.

Smartphone
Mobile phone built on a mobile computing platform that offers more advanced computing ability and connectivity than a standard mobile phone.

Software-as-a-service (SaaS)
The delivery of software applications remotely by a provider over the internet; perhaps through a web interface.

Spyware
Malware that passes information about a computer user's activities to an external party.

Supply chain
A set of organisations with linked resources and processes involved in the production of a product.

Tablet
An ultra-portable, touch screen computer that shares much of the functionality and operating system of smartphones, but generally has greater computing power.

Threat
Something that could cause harm to a system or organization.

Threat actor
A person who performs a cyber attack or causes an accident.

Two-factor
Obtaining evidence of identity by two independent means.

User name
The short name, usually meaningful in some way, associated with a particular computer user.

User account
The record of a user kept by a computer to control their access to files and programs.

Virtual private network (VPN)
Link(s) between computers or local area networks across different locations using a wide area network that cannot access or be accessed by other users of the wide area network.

Virus
Malware that is loaded onto a computer and then run without the user's knowledge or knowledge of its full effects.

Vulnerability
A flaw or weakness that can be used to attack a system or organization.

Wide area network (WAN)
Communications network linking computers or local area networks across different locations.

Wi-Fi
Wireless local area network based upon IEEE 802.11 standards.

Worm
Malware that replicates itself so it can spread to infiltrate other computers.

Index

Active cyber security, 9
Android devices, 101
Anti-spyware, 35
Antivirus protection, 9
Anti-virus software, 33
Apple iOS, 101
Attachments, 34, 36
Attacker, 11
Automatic scans, 36
Avast, 102
AVG, 102

Backup, 15, 19, 35
Basic router option, 72
Bring Your Own Device Policy, 98
Brute force attacks, 54

Circuit-level gateways, 77
Cloud storage, 21
Covid 19, 3
Cybercrime trends, 2

Index

Default usernames, 14
Denial of service attack, 3
Dictionary, 54

Email, 4, 15, 45
Europol, 3
Event based backup, 26

Firewall router, 73
Firewalls, 14, 35, 65

Google Android, 101

Hacker, 4, 21,
Hardware based firewalls, 66
Homeworking, 2

Identity theft, 55
Incremental backup, 20

Load balancer, 59, 74
Local Area Network (LAN), 21

Malicious code, 11, 33
Malware, 11, 33
Managed backups, 23, 25
Manual scans, 36

Index

McAfee, 102
Medical equipment, 4
Medical records, 4
Mobile phones, 97
Mobile security software, 98, 101
Multi-factor authentication, 14, 90

National Cyber Security Centre, 27
Network firewalls, 66

Online backup systems, 21
Open File Backups, 22

Packet filtering, 78
Passive security, 9
Passwords, 13, 35, 53
Password managers, 57, 59
Password overload, 58
Penetration testing, 9
Personal identification numbers (PINS), 53
Phishing, 9, 85
Polymorphic viruses, 13
Pop-up advertisements, 34
Portable Devices, 8, 113, 123
Privacy advisor, 102
Proxy deep packet inspection, 80
Public Wi-Fi, 36, 59

Index

Ransom attacks, 3
Rapid fire attacks, 4
Ransomware, 2, 21
Remote backup, 23
Remote photo, 104
Remote wipe, 104

Scheduled Backup, 26
Smartphone, 4, 97
Smishing, 8, 87
Social engineering, 54, 85
Software patches, 12, 35, 48
Stateful inspection firewall, 79
Stateless firewall, 79
Suspicious attachments, 89

Tablet, 4, 97
TCP handshake, 77
Third-party software firewall, 68, 70
Transmission control protocol, 72
Trojan horses, 21
Two factor authentication, 14

Unified threat management, 69, 75
Uniform resource locator (URL), 90
Unmanaged backups, 23

Index

Vishing, 86
Vulnerabilities, 12

Windows Defender, 68, 69

Other titles in the Straightforward Series

www.straightforwardco.co.uk

All titles, listed below, in the Straightforward Guides Series can be purchased online, using credit card or other forms of payment by going to www.straightfowardco.co.uk A discount of 25% per title is offered with online purchases.

Law

A Straightforward Guide to:

Consumer Rights
Bankruptcy Insolvency and the Law
Employment Law
Private Tenants Rights
Civil Justice After COVID
Family law
Small Claims in the County Court
Contract law
Intellectual Property and the law
Divorce and the law
Leaseholders Rights

The Process of Conveyancing
Knowing Your Rights and Using the Courts
Producing Your own Will
Housing Rights
Bailiffs and the law
Probate and The Law
Company law
What to Expect When You Go to Court
Give me Your Money-Guide to Effective Debt Collection
Caring for a Disabled Child

General titles

Letting Property for Profit
Buying, Selling and Renting property
Buying a Home in England and France
Bookkeeping and Accounts for Small Business
Creative Writing
Freelance Writing
Writing Your own Life Story
Writing performance Poetry
Writing Romantic Fiction
Speech Writing
Creating a Successful Commercial Website

The Straightforward Business Plan
The Straightforward C.V.
Successful Public Speaking
Starting an online business
Handling Bereavement
Individual and Personal Finance
Understanding Mental Illness

Crime Reference

The Crime Writers casebook
Being a Detective
A Comprehensive Guide to Arrest and Detention
A Comprehensive Guide to Burglary and Robbery
A Comprehensive Guide to Drink and Disorder

Go to: www.straightforwardco.co.uk